From The Stable Of The Platform of the

Flavor Of Mathematics

FLAVOUR

OF

MATHEMATICS

Concentrate on the MENSURATION

THE PLANE SHAPES

Volume 1

TEMITOPE JAMES

Author and mathematician
C.E.O: Flavor Of Mathematics
www.flavorofmathematics.com
temitopejames922@gmail.com

Mathematics is your food...........

FLAVOR OF MATHEMATICS
Is a

mathematical company which will attend to your mathematical needs at anytime.

For correspondence, issues, comments, invitations, media and business branding connect....email us and we'll respond to you within 24hours. Email @

info@flavorofmathematics.com

Or

Go through our website to know more about us on

www.flavorofmathematics.com

or visit our blog for mathematical updates, mathematical news, next coming up books from the FLAVOR OF MATHEMATICS @

flavorofmathematics.blogspot.com

ACKNOWLEDGEMENT

I want to appreciate my loved ones, mathematics colleagues and friends for showing me care, love, support and affection towards the publication of this book. You all will forever remain cherished in my heart.

All followers and friends on the Facebook page, flavor folks, our official websites, twitter and blog of the flavor of mathematics. Thanks for your encouragement, e – mails and support. I thank and appreciate you all for your comments.

DEDICATION

I dedicate this book to the Almighty God, for giving me the wisdom and knowledge to write this book with ease. May his name be praise forever.

I also dedicate this book to my lovely daughter (Esther James) and son (flavor James). The smile on your faces gives me joy to always appreciate your presence in my life.

I also dedicate this book to every devoted mathematician who took their time to contribute to the success of MATHEMATICAL EDUCATION all over the world.

PREFACE

The FLAVOR OF MATHEMATICS is an international mathematical company meant to inculcate the study of fundamental and basic mathematics into the lives of all clients and students of this new generation. Our duty is to create questions for you to solve comfortably in order to prepare you for every internal and external examination in the world.

From the stable of the FLAVOR OF MATHEMATICS, here comes the Most clearly stated sub topic workbook of the new generation; titled "FLAVOR OF MATHEMATICS" (Concentrate on the MENSURATION).

This book vividly explains the root of the MENSURATION in mathematics. It is a book that will expose every fact you need to know on the mensuration. It is a comprehensive explanatory book meant for all level of students to increase their intellectual quotient on the study of the MENSURATION for success in their mathematical examinations. This is a comprehensive textbook/journal that has a great advantage which is stated below;

- *Clearly stated explanatory at the beginning of every headlines and topics for proper and simple understanding*
- *Breakdown of Questions for understanding and learning for easy catch up of calculations and expressions*
- *The book contains challenging questions and options for gifted students of the world*
- *It contains comprehensive revision of questions for exercise*

- *Numerous graded questions to increase the intellectual quotient of the students*
- *Answers or hints to the questions for easy understanding and to grab each questions with ease to balanced studies*

This book is written, arranged, well typed students who are willing to be sound, brave and psychologically strong in mathematics. This book is useful for students of all categories in the WORLD dealing with business management mathematics, industrial mathematics, domestic mathematics and commercial mathematics. It is a book meant for both the junior and senior high school of all categories in the world, and also university level of advancement for the enlargement of mathematical studies for reference purposes.

Therefore, FLAVOR OF MATHEMATICS is a book recommended for you. Purchase this book, perceive the aroma, eat it, feed on it, wine and dine with it and strictly solve it. This book covers every explanation you need on the MENSURATION which will strengthen you for your breakthrough in mathematics.

From the FLAVOR OF MATHEMATICS....We beseeches you to have a copy of this book, sit and relax and enjoy the FLAVOR OF MATHEMATICS (Concentrate on the MENSURATION)..... Have a happy solving and reading.

TEMITOPE JAMES

C.E.O of the Flavor of Mathematics

OUR MESSAGE TO ALL STUDENTS AND PEOPLE

FROM THE

FLAVOR OF MATHEMATICS

M = *Many people dislike me because they feel I am too difficult*

A = *All shall be incomplete without me*

T = *Try to practice me and you shall get used to me*

H = *How sad some people feel when they hear of me*

E = *Employ me and find out that I am unique among all other courses*

M = *Many set solutions to their mathematical problems through me*

A = *At least, I help those who work with me*

T = *Try me and you shall be great among equals*

I = *It will be good for you if you concentrate on me*

C = *Come to me and you will be good in all calculations*

S = *Study me and you will realize I am not as difficult as you think.*

THE MENSURATION

The mensuration deals with the study of different shapes. These shapes are used for the calculation of different sizes of different shapes. The shapes and there formula are a case study in mathematics under the topic called the mensuration.

There are two types of shapes. They are

    i.   The plane shapes
    ii.   The solid shapes

- *The plane shapes deals on the study of the triangle, rectangle, square, circle, trapezium etc*
- *The solid shapes deals on the study of the cuboid, cube, cone, sphere, pyramid etc*

In this volume one of the mensuration, we shall deal on the study of the plane shapes. Students should concentrate on this topic which is the most common topic in mathematics dealing with the use of popular shapes

Different formulae are named after each shape which is compulsory for every student to know offhand for learning and reference studies. The plane shapes we are going to study are listed and explain briefly

i. **The triangles**
ii. **The rectangles**
iii. **The parallelogram**
iv. **The kite**
v. **The trapezium**
vi. **The circle**
vii. **The square**

The brief explanations of the above planes shapes are given with few samples to learn from their calculation

THE TRIANGLES

The triangle is a shape that is bounded with the sum of the three sides. The area of the triangle is equal to half of the product of its base and height. The diagram is given below with its formula

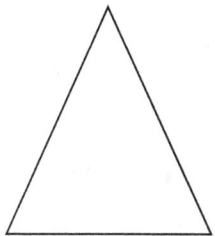

Formula for the triangles is;

i. Perimeter; $A + B + C$

ii. Area; when $X = \frac{1}{2}(a + b + c)$
Then area is $\sqrt{x(x-a)(x-b)(x-c)}$

iii. Area of the triangle give a angle with two sides = $\frac{1}{2}(ab)\sin\theta$

iv. Area of the triangle given a perpendicular height = $\frac{1}{2}bh$. When b is $(x + x)$

The samples and its solution of the triangle are given at the next page

Examples

1. Find the perimeter of the below triangle

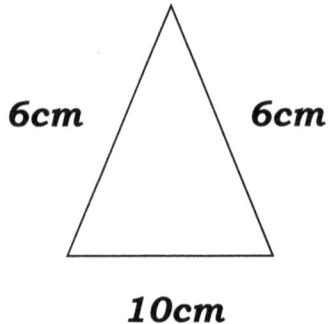

Solution

The perimeter of the triangle is A + B + C

   Where A = 6, B = 6 and C = 10
   It becomes 6 + 6 + 10 = 22cm
   The perimeter is 18cm

2. Find the perimeter of the below triangle

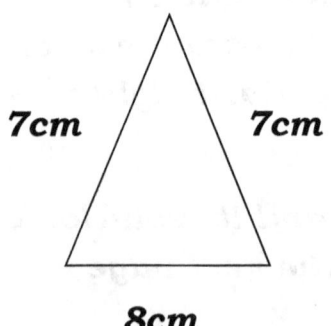

Solution

The perimeter of the triangle is A + B + C

Where A = 7, B = 7 and C = 8
It becomes 7 + 7 + 8 = 22cm

The perimeter is 22cm

3. When the perimeter of the below triangle is 30cm, what is the unknown value of the base of the triangle?

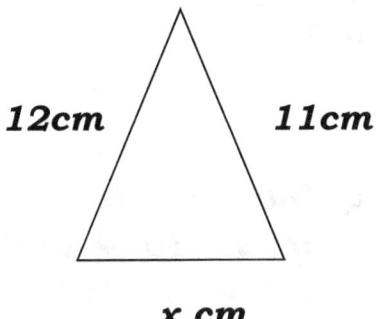

12cm   11cm

x cm

Solution

The perimeter of the triangle is A + B + C

Where A = 12, B = 11 and C = x
It becomes 12 + 11 + x = 30

The unknown value is 7cm

4. Find the areas of the triangles with the following dimension
   i. $a = 9cm$, $b = 16cm$ and $c = 12cm$
   ii. $a = 4cm$, $b = 10cm$ and $c = 12cm$

Solution

- **The area of the triangle is $X = \frac{1}{2}(a + b + c)$**
  Then, $a = 9cm$, $b = 16cm$ and $c = 12cm$
  It becomes $\frac{1}{2}(9 + 16 + 12)$
  $\frac{1}{2}(37) = 18.5cm$
  The area becomes
  $\sqrt{18.5(18.5 - 9)(18.5 - 16)(18.5 - 12)}$
  $\sqrt{18.5 \times 9.5 \times 2.5 \times 6.5}$
  $\sqrt{2855.9375}$
  $53.4cm^2$

- **The area of the triangle is $X = \frac{1}{2}(a + b + c)$**
  Then, $a = 4cm$, $b = 10cm$ and $c = 12cm$
  It becomes $\frac{1}{2}(4 + 10 + 12)$
  $\frac{1}{2}(26) = 13cm$
  The area becomes
  $\sqrt{13(13 - 4)(13 - 10)(13 - 12)}$
  $\sqrt{13 \times 9 \times 3 \times 1}$
  $\sqrt{351}$
  $18.7cm^2$

5. find the area of the below triangle

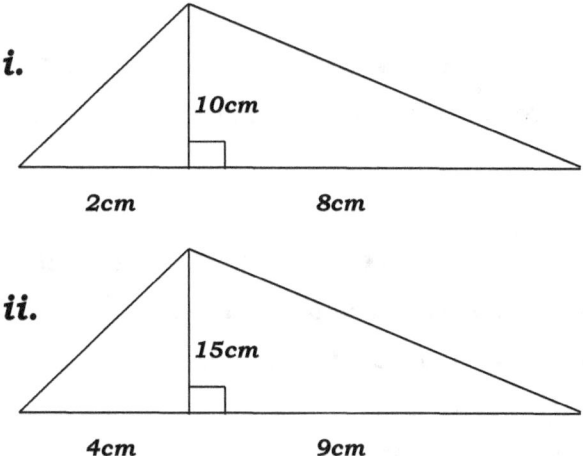

Solution

- The area of the triangle is ½ bh
  Where h = 10 and b = (2 + 8) = 10
  It becomes ½ (10)(10) = 50cm²

- The area of the triangle is ½ bh
  Where h = 15 and b = (4 + 9) = 13
  It becomes ½ (15)(13) = 97.5cm²

6. Find the area of a triangle with angle 52⁰ when a and b are given as 6cm and 3cm

Solution

The formula for the triangle when an angle is included is known as ½absin θ
Where a = 6cm and b = 3cm
½ (6 × 3) × sin 52⁰

It becomes 7cm²

7. What is the area of a triangle with angle 40^0 when x and y are given as 9cm and 10cm?
Solution
The formula for the triangle when an angle is included is known as ½absin θ
Where x = 9cm and y = 10cm
½ (9 × 10) × sin 40^0
It becomes 28.9cm²

8. The area of a triangle is 5cm². When the triangle has two sides as a and b which are given as 3cm and 5cm, what is the angle of the triangle?
Solution
The formula for the triangle when an angle is included is known as ½absin θ
Where a = 3cm and b = 5cm
½ (3 × 5) × sin $θ^0$ = 5
Then; 15 sin θ = 5 × 2
Sin θ = $10/15$
Sin θ = 0.66666666
Then, $\sin^{-1}$ 0.6666666 = 41.8^0
Therefore, the angle of the triangle is 41.8^0

NOTE;** **when solving questions on the triangle

- ***The result of the perimeter on triangle results to (cm)***
- ***The result of the area of the triangle results to (cm²)***

THE RECTAGLES

The rectangle is a four sided shape that is bounded with four sides which are not equal to each other. Two sides opposite each other are equal but not all sides in common are equal to each other. See the below diagram

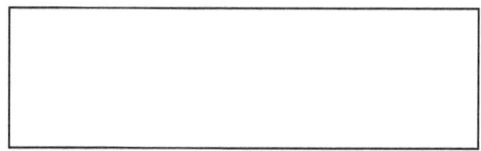

The formula of the rectangle are given below for reference purposes

i.   Area of a rectangle is (L × B)
ii. Perimeter of the rectangle is 2(L + B)
     Where L = Length and B = Breath

See samples below for reference studies of the rectangle

Examples

1. Find the area of the rectangle when the length is 6cm and the breath is 9cm
   Solution
   Formula = length × Breath
   Where l = 6 and b = 9

It becomes (6 × 9) = 54cm²

2. What is the area of the rectangle when the length is 10cm and the breath is 12cm?
Solution
Formula = length × Breath
Where l = 10 and b = 12
It becomes (6 × 9) = 54cm²

3. Find the breath of the rectangle when the length is 6cm and area of the rectangle is 42cm².
Solution
Formula = length × Breath
Where l = 6 and b = x
It becomes (6 × x) = 42cm²
It becomes x = $^{42}/_6$
The breath of the rectangle is 8cm

4. Find the perimeter of the rectangle when the length is 5cm and the breath is 3cm
Solution
Formula = 2(l + b)
Where l = 5 and b = 3
It becomes 2(5 + 3) = 16cm
Therefore, the perimeter of the rectangle is 16cm

5. What is the perimeter of the rectangle when the length is 15cm and the breath is 8cm?
Solution
Formula = 2(l + b)
Where l = 15 and b = 8
It becomes 2(15 + 8) = 46cm
Therefore, the perimeter of the rectangle is 46cm

6. Find the length of the rectangle when the perimeter of the rectangle is 30cm and the breath is 5cm
Solution
Formula = 2(l + b)
Where l = x and b = 5
It becomes 2(x + 5) = 30cm
2x + 10 = 30
2x = 30 − 10
2x = 20
x = 10
Therefore, the length of the rectangle is 10cm

NOTE; when solving questions on the rectangle

- The result of the perimeter on rectangle results to (cm)

- ***The result of the area of the rectangle results to (cm²)***

THE PARALELOGRAM

The parallelogram is divided into two congruent triangles. The area of the parallelogram is twice the area of one of these triangles while the perimeter is the addition of all the four sides. See the below diagram of the parallelogram

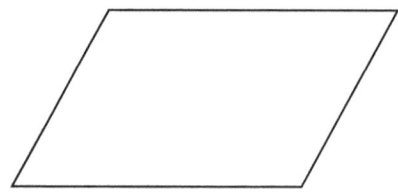

The formula of the parallelogram are given below

1. Area of the parallelogram; Base × Height
2. Perimeter of the parallelogram; 2(a + b)
3. Area of the parallelogram when the angle is included = ab sinθ

See the samples below for reference studies and purposes

1. Find the perimeter of the parallelogram in which the base is 6cm and the height is 9cm.

Solution
The perimeter is 2(a + b)
When a = 6cm and b = 9cm
2(6 + 9) = 30cm
The perimeter of the parallelogram is 30cm

2. **Find the perimeter of the parallelogram in which the base is 11cm and the height is 14cm.**
Solution
The perimeter is 2(a + b)
When a = 11cm and b = 14cm
2(11 + 14) = 50cm
The perimeter of the parallelogram is 50cm

3. **Find the area of the parallelogram in which the base is 5cm and the height is 7cm.**
Solution
The area is b × h
When b = 5cm and h = 7cm
(5 × 7) = 35cm²
The area of the parallelogram is 35cm²

4. **What is the area of the parallelogram in which the base is 10cm and the height is 6cm?**

Solution

The area is $b \times h$

When $b = 10cm$ and $h = 6cm$

$(10 \times 6) = 60cm^2$

The area of the parallelogram is $60cm^2$

5. **What is the area of the parallelogram in which the angle is 88^0, the base is 21cm and the height is 10cm?**

Solution

The area is $(ab \sin \theta)$

When $a = 21cm$ and $b = 10cm$

$(21 \times 10) \times \sin 88^0 = 209.8cm^2$

The area of the parallelogram is $209.8cm^2$

6. **What is the area of the parallelogram in which the angle is 60^0, the base is 14cm and the height is 8cm?**

Solution

The area is $(ab \sin \theta)$

When $a = 14cm$ and $b = 8cm$

$(14 \times 8) \times \sin 60^0 = 96.9cm^2$

The area of the parallelogram is 96.9cm^2

NOTE; when solving questions on the parallelogram

- *The result of the perimeter on parallelogram results to (cm)*
- *The result of the area of the parallelogram results to (cm^2)*

THE KITE

The kite is a four sided shape. The diagonals bisect at the right angle. When you look or observed the shape, it looks as if two triangles are merged together at their bases. The formula are given below

1. The perimeter; $2(a + b)$
2. Area; $Y(b_1 + b_2)$

Examples

Find the perimeter and the area of the below diagram

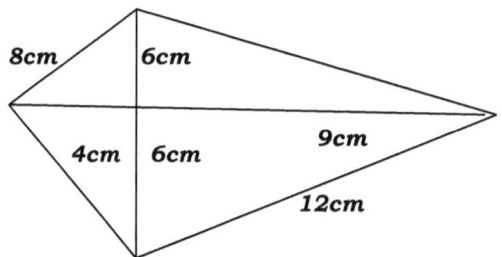

The perimeter of the kite; $2(l + b)$
Where $l = 8$ and $b = 12$
It becomes $2(8 + 12)$
   $40cm$

The area of the kite $Y(b_1 + b_2)$
   $6(4 + 9) = 78cm^2$

NOTE; *when solving questions on the kite*

- *The result of the perimeter on kite results to (cm)*
- *The result of the area of the kite results to (cm²)*

THE TRAPEZIUM

The shape of the trapezium is almost the same as the parallelogram but there is a slight difference in the diagram. See the below diagram.

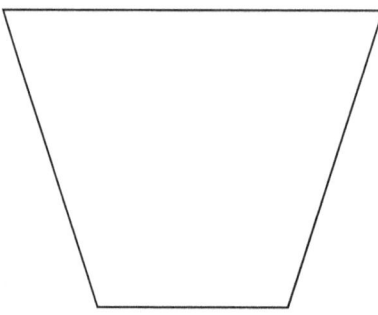

See the below formula of the trapezium
1. Area = ½ (a + b)h
2. Area included with an angle is ½ (a + b)c sin θ

Examples
1. Find the area of the trapezium if AB = 6cm, DC = 8cm, BC = 5cm with angle B at $60°$
Solution
Diagram given at the next page

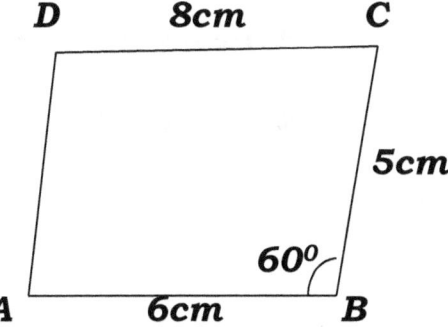

The formula for the trapezium when an angle is included is $60°$

½ (a + b) c sin θ

Where a = 8, c = 5 and b = 6, θ = $60°$

½ (8 + 6) × sin $60°$

30cm²

2. Find the area of the trapezium, if AB = 6cm, DC = 12cm and BC = 6cm

Solution

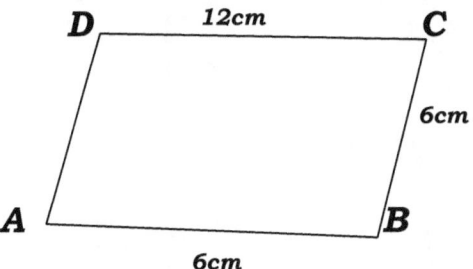

Area of the trapezium = ½ (a + b) h

½ (12 + 6) 6

54cm²

NOTE; when solving questions on the trapezium

- ***The result of the area of the kite results to (cm²)***

THE CIRCLE

The circle is a plane shape bounded by the curve surface. It is one of the most popular plane shapes in mathematics used for different purposes. The diagram is seen below

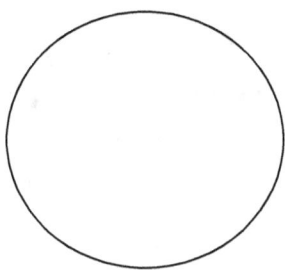

The circular part of the circle is known as the circumference. The perimeter of the circle is known as the circumference of a circle. See the below formula of the circle

i. *The circumference of a circle; $2\pi r$*
ii. *The area of a circle; πr^2*
   See the below samples of the circle

Examples
1. *Calculate the circumference and the area of a circle of radius 14cm*
   Solution
   The formula of the circumference of the circle is $2\pi r$ where $\pi = {}^{22}/_7$ and $r = 14$

It becomes $2 \times {}^{22}/_7 \times 14$
= 88cm

2. Calculate the circumference and the area of a circle of radius 20cm
Solution
The formula of the circumference of the circle is $2\pi r$ where $\pi = {}^{22}/_7$ and $r = 20$
It becomes $2 \times {}^{22}/_7 \times 20$
= 125.7cm

3. Find the area of the circle when the radius of the circle is 9cm
Solution
The formula for the area of the circle is πr where $\pi = {}^{22}/_7$ and $r = 9$cm
πr^2 will become ${}^{22}/_7 \times 9^2$
${}^{22}/_7 \times 9 \times 9 = 254.57 cm^2$

4. Find the area of the circle when the radius of the circle is 13cm
Solution
The formula for the area of the circle is πr where $\pi = {}^{22}/_7$ and $r = 13$cm
πr^2 will become ${}^{22}/_7 \times 13^2$
${}^{22}/_7 \times 13 \times 13 = 531.14 cm^2$

5. When the circumference of the circle is 120cm, what is the radius of the circle?

Solution

The formula of the circumference of the circle is $2\pi r$ where $\pi = {}^{22}/_7$ and $r = x$, and circumference of the circle is 120

It becomes $2 \times {}^{22}/_7 \times x = 120$

$= 19.09cm$

The radius if the circle is 19.09cm

NOTE; when solving questions on the circle

- **The result of the circumference of the circle results to (cm)**
- **The result of the area of a circle results to (cm^2)**

THE SQUARE

The square is a shape that is bound by four equal sides. The square is a shaped used in our domestic area of life for different purpose and it is also used for many calculations in mathematics. The below diagram is a square.

The formula for the square is given below

i. The perimeter of a square is $4l$
ii. The area of a square is l^2

See samples of the square below

Examples

1. What is the perimeter of the square of length 4cm?
   Solution
   The perimeter of the square is $4l$ where l is 4. Then, $4 \times 4 = 16cm$

2. **What is the perimeter of the square of length 10cm?**
   Solution
   The perimeter of the square is $4l$ where l is 10. Then, $4 \times 10 = 40cm$

3. **Find the area of the square of length 8cm?**
   Solution
   The area of the square is l^2 where l is 8. Then, $8^2 = 64cm$

4. **Find the area of the square of length 12cm?**
   Solution
   The area of the square is l^2 where l is 12. Then, $12^2 = 144cm$

The plane shapes have been briefly explained with their samples in order to increase our reference studies of the plane shapes. All students and tutors should purchase the international mathematical workbook titled "mathematics is your food" (The food of mensuration volume 1 and 2) to solve more questions on the planes shapes and the solid shapes. Students and tutors must know the formula attached to the plane shapes in order to avoid error in their studies. The summary of the study of the mensuration is given below.

THE FORMULA SUMMARY

i. Perimeter; $A + B + C$
Area; when $X = \frac{1}{2}(a + b + c)$
The area is $\sqrt{x(x-a)(x-b)(x-c)}$
Area of the triangle give a angle with two sides = $\frac{1}{2}(ab)\sin \theta$
Area of the triangle given a perpendicular height = $\frac{1}{2}bh$. When b is $(x + y)$

ii. TRAPEZIUM; Area = $\frac{1}{2}(a + b)h$
Area include with an angle = $\frac{1}{2}(a + b)c \sin \theta$

iii. PARALELOGRAM; perimeter = $2(A + B)$
Area = AB
Area included with an angle = $AB\sin\theta$

iv. KITE; Perimeter = $2(A + B)$
Area = $x(A_1 + A_2)$

v. CIRCLE; Circumference = $2\pi r$
Area = πr^2

vi.   RECTANGLE
   Area = length × breath
   Perimeter = 2(A + B)

vii.   SQUARE;   Area = L^2
   Perimeter = 4L

The next volume of the mensuration contains the advance study of the plane shapes consisting of the arcs, chords, sectors and the segments. It is very important as a student or tutor to have a copy of the book from the platform of the flavor of mathematics.

The class exercise of the above planes shapes are given below to increase our intellectual quotient of studies. Solve the questions below and enjoy the flavor of mathematics

Class exercise

1. When a = 4cm, b = 6cm and c = 8cm, find the area and the perimeter of the triangle

2. Find the area of the triangle when a = 4cm, b = 6cm and c = 98⁰

3. Find the area of the perimeter of a rectangle of length 12cm and the breath of 10cm

4. Find the perimeter and the area of a parallelogram when the base 7cm and the height is 20cm

5. Find the area of the parallelogram when the base is 6cm , height is 8cm with an angle of 120⁰

6. Find the area and the perimeter of the below diagram

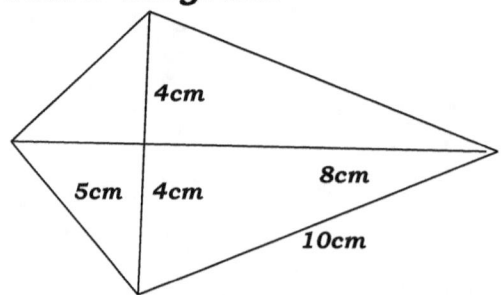

7. Find the area of the trapezium when PQ = 4cm, RS = 6cm ad QS = 8cm in the below diagram.

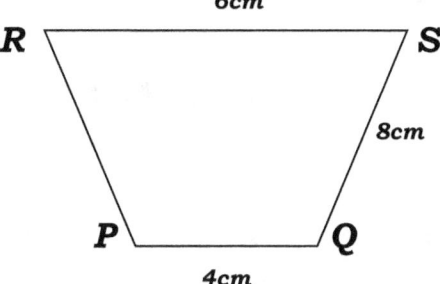

Also, find the area of the trapezium when the angle $34°$ is included into the above diagram

8. What is the circumference of the circle with the radius 8cm?

9. Find the circumference of the circle of radius 16cm.

10.   What is the area of the circle of radius 7cm?

11.   Find the area of the circle of radius 4cm

12. Find the perimeter of the square when the length is 6cm

13. What is the perimeter of the square when its length is 12cm?

14. Find the area of the square when its length is 13cm.

15. What is the area of the square when its length is 10cm?

Answers

1. Area = $11.6cm^2$, perimeter = $18cm$

2. $12cm^2$

3. Area = $120cm^2$, perimeter = $44cm$

4. Perimeter = $54cm$, Area = $140cm$

5. $41.56cm^2$

6. Area = $52cm^2$, perimeter = $38cm$

7. $40cm^2$ and $22cm^2$

8. $50.2cm$

9. $100.5cm$

10. $154cm^2$

11. $50.2cm^2$

12. $24cm$

13. $48cm$

14. $169 cm^2$

15. $100 cm^2$

From **The** *Flavor Of* **Mathematics**, our concern is for you to be academically brilliant on the study of mathematics. Our books are meant for the beginners (The kids), the basic level, the college level and also students from the higher institution can make use of our books. Visit our website for you to know where you can purchase our listed books. The flavor of mathematics is really concerned about you.

i. How are you coping on the study of mathematics?
ii. Are you scared of the branches of mathematics?
iii. Do you need books as a reference purpose to study more questions from any basic topic in mathematics?
iv. Is mathematics a problem in your academic life?
v. Do you need intellectual revival on the study of mathematics?

Then.... *Flavor Of* **Mathematics** **is the answer to all your questions. Purchase our books and break free from the Oppression of Mathematics. 100% SUCCESS is GUARANTEED to you when you are connected to us. See the our listed books below and have great day ahead**

Our books from the
Flavor Of Mathematics

1. *Flavor of Mathematics volume 1*
2. *Flavor of mathematics volume 2*
3. *James Application of Equations volume 1*
4. *James Application of Equations volume 2*
5. *James Application of Equations volume 3*
6. *My formulae (volume 1)*
7. *My formulae (volume 2)*
8. *Mathematics is your food ..(Algebraic process volume 1)*
9. *Mathematics is your food ..(Algebraic process volume 2)*
10. *Mathematics is your food ..(Alpha and beta volume 1)*
11. *Mathematics is your food ..(Alpha and beta volume 2)*
12. *Mathematics is your food ..(Arithmetic processes volume 1)*
13. *Mathematics is your food ..(Arithmetic processes volume 2)*
14. *Mathematics is your food ..(Arithmetic progression volume 1)*
15. *Mathematics is your food ..(Arithmetic progression volume 2)*
16. *Mathematics is your food ..(Binomial theorem volume 1)*

17. Mathematics is your food ..(Binomial theorem volume 2)
18. Mathematics is your food ..(Change of subject formula volume 1)
19. Mathematics is your food ..(Conic section volume 1)
20. Mathematics is your food ..(Conic section volume 2)
21. Mathematics is your food ..(Change of subject formula volume 2)
22. Mathematics is your food ..(Decimal volume 1)
23. Mathematics is your food ..(Decimal volume 2)
24. Mathematics is your food ..(Differentiation volume 1)
25. Mathematics is your food ..(Decimal volume 2)
26. Mathematics is your food..(Degree and Radian) Volume 1
27. Mathematics is your food..(Degree and Radian) Volume 2
28. Mathematics is your food ..(Elevation and Depression volume 1)
29. Mathematics is your food ..(Elevation and Depression volume 2)
30. Mathematics is your food ..(Simultaneous Equations volume 1)
31. Mathematics is your food ..(Simultaneous Equations volume 2)
32. Mathematics is your food ..(Significant figures Equations volume 1)

33. Mathematics is your food ..(Significant figures Equations volume 2)
34. Mathematics is your food ..(Factorization volume 1)
35. Mathematics is your food ..(Factorization volume 2)
36. Mathematics is your food ..(Fraction theories volume 1)
37. Mathematics is your food ..(Fraction theories volume 2)
38. Mathematics is your food ..(Geometric progression volume 1)
39. Mathematics is your food ..(Geometric progression volume 2)
40. Mathematics is your food ..(Indices volume 1)
41. Mathematics is your food ..(Indices volume 2)
42. Mathematics is your food ..(Inequalities volume 1)
43. Mathematics is your food ..(Inequalities volume 2)
44. Mathematics is your food ..(Integration volume 1)
45. Mathematics is your food ..(Integration volume 2)
46. Mathematics is your food ..(Longitude and Latitude volume 1)
47. Mathematics is your food ..(Longitude and Latitude volume 2)
48. Mathematics is your food ..(Matrixes volume 1)
49. Mathematics is your food ..(Matrixes volume 2)

50. *Mathematics is your food ..(Mensuration volume 1)*
51. *Mathematics is your food ..(Mensuration volume 2)*
52. *Mathematics is your food ..(Number base volume 1)*
53. *Mathematics is your food ..(Number base volume 2)*
54. *Mathematics is your food ..(Number substitution volume 1)*
55. *Mathematics is your food ..(Number substitution volume 2)*
56. *Mathematics is your food ..(logarithm volume 1)*
57. *Mathematics is your food ..(logarithm volume 2)*
58. *Mathematics is your food ..(logic volume 1)*
59. *Mathematics is your food ..(logic volume 2)*
60. *Mathematics is your food ..(Statistics volume 1)*
61. *Mathematics is your food ..(Statistics volume 2)*
62. *Mathematics is your food ..(Straight line of the Geometry volume 1)*
63. *Mathematics is your food ..(Straight line of the Geometry volume 2)*
64. *Mathematics is your food ..(Surds volume 1)*
65. *Mathematics is your food ..(Surds volume 2)*
66. *Mathematics is your food ..(Trigonometry volume 1)*

67. **Mathematics is your food ..(Trigonometry volume 2)**
68. **Mathematics is your food ..(Variation volume 1)**
69. **Mathematics is your food ..(Variation volume 2)**
70. **Mathematics is your food ..(Parallel line Geometry volume 1)**
71. **Mathematics is your food ..(Parallel line Geometry volume 2)**
72. **Mathematics is your food ..(Partial fraction volume 1)**
73. **Mathematics is your food ..(Partial fraction volume 2)**
74. **Mathematics is your food ..(Permutation and combination volume 1)**
75. **Mathematics is your food ..(Permutation and combination volume 2)**
76. **Mathematics is your food ..(Polygon volume 1)**
77. **Mathematics is your food ..(Polygon volume 2)**
78. **Mathematics is your food ..(Polynomials volume 1)**
79. **Mathematics is your food ..(Polynomials volume 2)**
80. **Mathematics is your food ..(Probability volume 1)**
81. **Mathematics is your food ..(Probability volume 2)**
82. **Mathematics is your food ..(Quadratic equations volume 1)**

83. Mathematics is your food ..(Quadratic equations volume 2)
84. Mathematics is your food ..(Roman figures volume 1)
85. Mathematics is your food ..(Roman figures volume 2)
86. Mathematics is your food ..(Sequence and Series volume 1)
87. Mathematics is your food ..(Sequence and Series volume 2)
88. Mathematics is your food ..(Set theory volume 1)
89. Mathematics is your food ..(Set theory volume 2)
90. Mathematics is your food ..(Standard form volume 1)
91. Mathematics is your food ..(Standard form volume 2)
92. Mathematics is your food ..(The Circle theorem volume 1)
93. Mathematics is your food ..(The Circle theorem volume 1)
94. James conversion volume 1
95. James conversion volume 2
96. James conversion volume 3
97. James conversion volume 4
98. James conversion volume 5
99. James conversion volume 6
100. James conversion volume 7
101. James conversion volume 8
102. James step 3, 2, 1 equation volume 1
103. James step 3, 2, 1 equation volume 2
104. James step 3, 2, 3 equation volume 3
105. James step 3, 2, 3 equation volume 4
106. James step 3, 2, 4 equation volume 5

107. James step 3, 2, 4 equation volume 6
108. James step 4, 3, 4 equation volume 7
109. James step 4, 3, 4 equation volume 8
110. James step 4, 3, 2, 1 equation volume 9
111. James step 4, 3, 2, 1 equation volume 10
112. James step 5, 4, 3, 2, 1 equation volume 11
113. James step 5, 4, 3, 2, 1 equation volume 12
114. James Application of equations on Matrix volume 1
115. James Application of equations on Matrix volume 2
116. James Application of equations on Surds volume 1
117. James Application of equations on Surds volume 2
118. James Application of equations (Questions) volume 1
119. James Application of equations (Questions) volume 2
120. Concentrate on the algebraic processes.. volume 1
121. Concentrate on the algebraic processes.. volume 2
122. Concentrate on the algebraic processes.. volume 3
123. Concentrate on the algebraic processes.. volume 4
124. Concentrate on the algebraic processes.. volume 5
125. Concentrate on the Approximation and Estimation volume 1

126. Concentrate on the Approximation and Estimation volume 2
127. Concentrate on the Arithmetic Progression volume 1
128. Concentrate on the Arithmetic Progression volume 2
129. Concentrate on the Alpha and Beta theory
130. Concentrate on the Binomial theorem
131. Concentrate on the Binary System
132. Concentrate on the Conic Section volume 1
133. Concentrate on the Conic Section volume 2
134. Concentrate on the Calculus volume 1
135. Concentrate on the Calculus volume 2
136. Concentrate on the Circle Geometry
137. Concentrate on the Geometric Progression volume 1
138. Concentrate on the Geometric Progression volume 2
139. Concentrate of the James Application of equations on matrix
140. Concentrate on the James Conversion volume 1
141. Concentrate on the James Conversion volume 2
142. Concentrate on the James Conversion volume 3
143. Concentrate on the Logic
144. Concentrate on the Longitude and Latitude
145. Concentrate on the Law of Indices
146. Concentrate on the Logarithm volume 1

147. *Concentrate on the Logarithm volume 2*
148. *Concentrate on the Matrixes volume 1*
149. *Concentrate on the Matrixes volume 2*
150. *Concentrate on the Mensuration volume 1*
151. *Concentrate on the Mensuration volume 2*
152. *Concentrate on the Mensuration volume 3*
153. *Concentrate on the Parallel line Geometry*
154. *Concentrate on the Partial Fraction*
155. *Concentrate on the Permutation and combination*
156. *Concentrate on the Polygon*
157. *Concentrate on the Polynomials volume 1*
158. *Concentrate on the Polynomials volume 2*
159. *Concentrate on the Probability*
160. *Concentrate on the Relations and Functions*
161. *Concentrate on the Sequence and Series*
162. *Concentrate on the Set theory volume 1*
163. *Concentrate on the Set theory volume 2*
164. *Concentrate on the Statistics volume 1*
165. *Concentrate on the Statistics volume 2*
166. *Concentrate on the Statistics volume 3*

167. Concentrate on the Statistics volume 4
168. Concentrate on the Statistics volume 5
169. Concentrate on the Statistics volume 6
170. Concentrate on the straight line of Geometry
171. Concentrate on the Surds
172. Concentrate on the Trigonometry 1
173. Concentrate on the Trigonometry 2
174. Concentrate on the Trigonometry 3
175. Concentrate on the Compound Interest
176. Concentrate on the Percentage
177. Concentrate on the Ratio
178. Concentrate on the Simple Interest
179. Concentrate on the Variation
180. Concentrate on the James application on equation on Surd
181. Standing on the binomial theorem
182. Standing on the Mathematical tables volume 1
183. Standing on the Mathematical tables volume 2
184. Standing on the Roman figure volume 1
185. Standing on the Roman figure volume 2
186. Mathematics for Beginners volume 1 (Counting of Numbers)
187. Mathematics for Beginners volume 2 (Counting of Numbers with shapes)
188. Mathematics for Beginners volume 3 (Filling the gaps with Numbers)

189. Mathematics for Beginners volume 4
(The learning of Fractions)
190. Mathematics for Beginners volume 5
(The even Numbers)
191. Mathematics for Beginners volume 6
(The even Numbers...Practical)
192. Mathematics for Beginners volume 7
(The odd Numbers)
193. Mathematics for Beginners volume 8
(The odd Numbers....Practical)
194. Mathematics for Beginners volume 9
(The Prime Numbers)
195. Mathematics for Beginners volume 10
(The Prime Numbers....Practical)
196. Mathematics for Beginners volume 11
(The Time Clock)
197. Mathematics for Beginners volume 12
(The Time Clock....Practical)
198. Mathematics for Beginners volume 13
(Number in words)
199. Mathematics for Beginners volume 14
(Number in words....Practical)
200. Mathematics for Beginners volume 15
(The addition)
201. Mathematics for Beginners volume 16
(The addition....Practical)
202. Mathematics for Beginners volume 17
(The Subtraction)
203. Mathematics for Beginners volume 18
(The Subtraction....Practical)
204. Mathematics for Beginners volume 19
(The Multiplication)
205. Mathematics for Beginners volume 20
(The Multiplication....Practical)

206. *Mathematics for Beginners volume 21 (The Division)*
207. *Mathematics for Beginners volume 22 (The Division….Practical)*
208. *Mathematics for Beginners volume 23 (The Square root and Cube root)*

More books from us are on the way to educate you on the study of mathematics. We are committed, reliable, strong and efficient to increase your intellectual quotient of the study of mathematics. Stay glue to our website and blog for more information. Have a wonderful time with us

IMPORTANT NOTICE TO OUR ADOURABLE STUDENTS, TEACHERS AND CLIENTS

We apologise for any topographical error in this book. The FLAVOR OF MATHEMATICS is an English owned mathematics company. The mistakes are NOT intentional and we will try our best to aknowledge your emails and comments on our website. Also, we kindly urge you to accept us and patronize our books to help you succeed in your mathematical careers. We are also happy to let you know that ALL OUR BOOKS ARE TRANSLATED INTO THE FOLLOWING LANGUAGES to suite your desired language and interest;

- *English Language*
- *Italian Language*
- *German Language*
- *French Language*
- *Spanish Language*

Depending on your favorite language, purchase the books, solve from it, learn from it inculcate from it and enjoy the FLAVOR OF MATHEMATICS.

www.ingramcontent.com/pod-product-compliance
Lightning Source LLC
Chambersburg PA
CBHW070406190526
45169CB00003B/1129